Juan Manuel Soto Parra
Rosa María Yáñez Muñoz
Nubia Guadalupe Torres Beltrán

Apple Fruit Immersion Effect

Juan Manuel Soto Parra
Rosa María Yáñez Muñoz
Nubia Guadalupe Torres Beltrán

Apple Fruit Immersion Effect

in Cation Solutions on Postharvest Quality

ScienciaScripts

Cover image: www.ingimage.com

This book is a translation from the original published under ISBN 978-613-9-40948-8.

Publisher:
Sciencia Scripts
is a trademark of
Dodo Books Indian Ocean Ltd. and OmniScriptum S.R.L publishing group

120 High Road, East Finchley, London, N2 9ED, United Kingdom
Str. Armeneasca 28/1, office 1, Chisinau MD-2012, Republic of Moldova, Europe
Printed at: see last page
ISBN: 978-620-7-88732-3

EFFECT OF APPLE FRUIT IMMERSION IN CATION SOLUTIONS ON POSTHARVEST QUALITY

JUAN MANUEL SOTO PARRA ROSA MARIA YÂNEZ MUNOZ
NUBIA GUADALUPE TORRES BELTRÂN

"Success is not about how much you accumulate for yourself, but how much you accumulate for yourself. how much yuec[es c[ar to {os c[emas"].

Norman 'Bor{augaug

INDEX

INTRODUCTION

Extending the shelf life of fruit is one of the major challenges for producers. Consumers' evaluation of apples is based on their appearance (colour, size, shape, absence of defects) and then on their eating quality (Musacchi and Serra, 2018). To preserve quality, reduce losses and extend the period of distribution and consumption, apples are subjected to cold storage, which aims to slow down the metabolism of the fruit after harvesting (Falcon, 2016).

Mineral elements such as cations play a specific role in crops, e.g. in shelf life. The presence of calcium ensures fruits with greater firmness (Valdiviezo, 2018), in addition, the immersion of fruits in calcium solutions has a very similar effect to that achieved with sprays, a treatment that has a positive effect on fruit cooling (Monge, et al, 1994), while, in young fruits, potassium is indispensable to obtain greater turgidity, which is required for cell expansion (Leon, 2016). On the other hand, magnesium is involved in energy uptake, transformation and transport (Blackhall, et al, 2018), and sodium contributes to osmosis (movement of water in and out of plant cells) (Cortés, 2019).

Golden Delicious' apples, due to their epicarp with thin cuticle properties and high presence of lenticels, are sensitive to water loss and wilting at different stages of ripening, being more susceptible in cold storage (Cepeda, et al, 2014).

In this sense, it is necessary to improve the apple production chain, since a good postharvest management generates a reduction in fruit loss. For this reason, the present research focuses on the contribution of a new postharvest scheme in the apple growing region, with the objective of implementing the application and balance of cations such as potassium, calcium, magnesium and sodium in 'Golden Delicious' apple trees by means of the fruit immersion technique.

LITERATURE REVIEW

General information about Manzano

The history of the apple is not known with certainty, according to a review by Tarrillo (2017) the wild species (malus sieversii (Ledeb) appeared more than 15,000 years ago in regions of Asia, domesticated by civilisations such as the Greeks, Romans and Egyptians, and it was not until the 15th century that the first crops were planted in America (Morales, 1999). Castillo (2018) reported that in Chihuahua the introduction of the crop was from the colonisation when Spanish evangelists arrived in the region, cultivating their own fruit orchards and teaching the settlers its management, after many years it became a crop of economic importance in the state.

The apple tree (malus domestica) is a fruit tree that is cultivated worldwide, it has more than 30 species and around 1,000 varieties that develop in temperate-cold climates (Posadas et al., 2018), it is a deciduous tree (Jackson and Palmer, 1999), generally self-sterile so it requires cross-pollination with other compatible varieties and the help of insects and wind. The fruit is composed of the epicarp (skin), the mesocarp (pulp), and the endocarp (core) (Zaera, 2008), the apple has characteristics of great importance in human nutrition due to the contribution of phenolic compounds and flavonoids, being antioxidants with a positive influence, widely documented, on health, mainly when consumed fresh (Palomo et al., 2010).

Taxonomy

The scientific name of the apple tree is: Malus domestica

Their scientific classification is shown in table 1.

Table 1 Classification of apple trees

Reino	Plantae
Division	Magnoliophyta
Order	Rosebushes
Family:	Rosaceae
Gender:	Malus

(Barraquilla, 2005)

A number of apple varieties exist and to date new ones are still being produced and worked on to improve the value of the fruit in the food industry as a strategy to numerous challenges mainly in plant material, production technology, post-harvest and brand development (Arellano, 2021).

Fertilisation

Nutrition in apple trees is a determining factor to obtain a quality production. The positive effect of the fertilizer on the fruit is influenced by the adequate balance of the elements and the way they are applied, since their excesses or deficiencies act directly on the composition of the fruit, affecting the harvest and postharvest (Ponce et al., 2020; Mancera et al., 2007).The macro and micro elements in apple fruit have a specific and direct function on its quality. In the nutritional content of the fruit, the major minerals (macroelements) required in large quantities are: nitrogen (N), phosphorus (P), potassium (K), calcium (Ca) and magnesium (Mg), while the microelements required in small quantities are: zinc (Zn), iron (Fe), copper (Cu), boron (B) and manganese (Mn). All at the same level of importance (Gutiérrez et al., 2014). For any crop it is necessary to carry out physical and chemical studies of the soils at least per year, in apple trees according to Soto et al., (2007) it should be considered that the extractions for the harvest are generally 60 units of N fertiliser, for P it is estimated 80 UF and for K 125 UF, making applications of low doses of microelements, to achieve a quality harvest. Furthermore, Mg, together with K and Ca, is one of the three fundamental cations of living matter, it is necessary for the functioning of enzymes, to produce

carbohydrates and sugars used for fruit (Valdivezo, 2017). As for K and Mg, they are involved in phloem loading and transport of sugars from source organs to sinks, their uptake by roots and transport of Mg by shoots in plants are inhibited by increased K supply, which may exert detrimental effects on productivity and nutritional quality of harvested products, however, there are inconsistent effects of Mg on K uptake and translocation, which probably depend on plant species and rates and ratios of K and Mg application in the growth medium, there are synergistic effects of Mg and K on photosynthesis, carbohydrate transport and allocation, N metabolism and cell osmotic balance. To avoid antagonistic interaction between K and Mg and its adverse consequences on yield and nutritional quality, special attention should be paid to K and Mg balanced nutrition (Xie et al., 2021). In most cases, when the K/Mg ratio supplied to plants is imbalanced, the relationship between K and Mg is often antagonistic in the source organs, but synergistic in the sinks (Koch et al., 2019).

Post-harvest

The period of time between fruit harvesting and consumption can lead to quality losses due to physical, chemical, enzymatic or microbiological changes and the consequences of quality loss are economic losses (Kyanko et al., 2010).

The autocatalytic production of ethylene provokes several physiological changes known as climacteric, causing an increase in the respiration of the fruit and as a consequence the organoleptic ripening of the fruit is generated, this metabolic process is given thanks to compounds such as sugars, starch and fatty acids, leading them to an oxidative degradation, which results in simpler molecules such as carbon dioxide (CO_2) and water (H_2O), also produces more complex molecules that are used in photosynthesis, in all this process energy is released in the form of ATP (Candan and Calvo, 2017). In the case of apple as a climacteric fruit, it has the characteristic of ethylene production; ripening hormone, on which several research works and technologies have been developed to inhibit its production in order to prolong the commercial life of fruits, some of these strategies can be mentioned as: the use of ethylene inhibitors (1- MCP) (Trainotti et al., 2007), application of gamma and ultraviolet radiation, application of edible coatings (Fernandez et al., 2015), the use of plastic packaging, the use of pellicles (Tovar et al., 2011) and fruit fertilisation such as immersions in nutrient solutions to the application of controlled atmospheres with very low oxygen levels, and refrigeration chambers, all used directly or indirectly to alter ethylene action, since the higher the respiration rate of a fruit, the shorter its shelf life, (Kumar et al., 2014).

Low temperature storage is the most efficient method to increase the shelf-life of fruits due to its effect on slowing down the ripening rate, increasing their storage life, reducing losses and extending the period of distribution and consumption (Candan and Calvo, 2017). However, during cold storage, the firmness of the pulp gradually decreases, there is a certain loss of aromas, the acid and sucrose content is reduced, there are weight losses due to transpiration and, under certain conditions, physiopathologies and rots appear (Falcon, 2016), such as, for example, long periods of time, the inadequate use of controlled atmosphere conditions on the type of product, leading to a loss of quality. Another important parameter to take care of is high temperature and humidity, as they favour the growth of superficial and internal moulds and the activity of infecting insects (Gomila, 2016), which leads to significant quality and, therefore, economic losses (Falcon, 2016).

Balance of Cations

The uptake of nutrients in the soil is directly related to the cation exchange capacity and the valence number of the ion involved in the process. Some ions such as K^+ and NH_4^+ require one H ion$^+$ for exchange, while Ca^{2+} and Mg^{2+} require two. An important aspect of cation exchange is that monovalent ions are more mobile than ions with a higher number of valences; additionally, ions with smaller diameters penetrate better than larger ions (Santos et al., 1999). On the other hand, the uptake of foliar cations and anions, which must penetrate the cuticle in equivalent amounts to maintain electrical neutrality, is of utmost importance, as cation retention is affected by several factors, Some cations are retained more strongly than others, generally due to the presence of moisture on the cuticle and the hygroscopicity of the salt. In order for the inorganic salts to penetrate without problem, the dissolution of the salt is necessary, which is determined by the deliquescence point (DDP). When the humidity is below the PDD, the salt remains in a solid state and cannot penetrate the cuticle, such is the case of KNO_3 and salts containing phosphorus, which are much less favoured, and on the contrary, if the PDD is above it, it can dissolve and enter easily, for example: $CaCl_2$ (33 %), $MgCl_2$ (33 %), K_2 CO_3 (44 %), Ca (NO)$_{32}$ (56 %) and Mg (NO)$_{32}$ (56 %) are salts feasible to penetrate the cufcula (Sanchez and Curetti, 2021).

According to Ruales (2019) a cation is an ion with a positive electric charge, i.e. it has lost electrons, in this case, Magnesium (Mg), Potassium (K), Calcium (Ca) and Sodium (Na), belong to the classification of cations, supporting the plant in:

- Efficient photosynthetic function.
- Nutrient translocators.
- They are involved in the translocation into energy and building substances.
- They reinforce cellular structures.

For example, the right cation balance can contribute to sustain the quality of apples, as Pavicié (1993) observed that K as an individual element can also affect the occurrence of bitter pit in apples. However, the interaction of elements such as K/Ca and Na/Ca, P/Ca and Mg/Ca, showed closer connection with bitter pit than the same elements applied individually (Ben, 1998). It is believed that high quality fruits tend to contain a high proportion of K and Ca and a low amount of Cu, Zn and Mn. Also, the occurrence of bitter pit is closely related to the amount of Ca and other mineral elements (Liu and Han, 1997). Ferguson and Watkins (2003), observed that the concentration and distribution of Ca, Mg and K in apples treated with Ca sprays during the growing season, and in fruit stored after post-harvest vacuum infiltration, showed highest concentrations of each cation in the skin and core and lowest concentrations in the outer rind. Ca sprays increased Ca levels in each tissue segment while K levels were lower in Ca-sprayed fruit, but effects on Mg were variable and concentrations of all 3 cations in the outer rind increased during storage.

Potassium

Potassium (K) is an essential element in high demand by crops for normal growth and development. Some of the main functions of potassium on plants are directed to osmoregulation, starch synthesis, enzyme activation, protein synthesis, stomatal movement, ionic charge balance (Hernandez, 2010), numerous biochemical and physiological processes vital for growth, yield, quality (Lester et al., 2010), resistance to adverse conditions such as cold or disease, also responsible for the translocation of nutrients and newly grown photosynthates to points of differentiation and contributes to fruit set (FAO, 2010). In the case of fruit quality, K stands out as the cation that has the greatest influence on organoleptic parameters such as the synthesis of polyphenols responsible for the colour and aroma of the fruit, which determines the marketability of the fruit, and also, through its mobility in the phloem and xylem, helps in the transport of solutes and the participation of assimilates,consumer-preferred qualities, as it provides phytonutrients of vital importance for human health (Lester et al., 2010; Tagliavini, 2015). For example, up to 50% of the total K absorbed by grapevines accumulates in berries, its functions in the fruit are related to synthesis reactions and enzyme activation, contributing directly to fruit ripening, sugar synthesis and maintenance of cell turgor. In mature and highly productive apple orchards, K is often the nutrient most in demand, apple, in fact, is a strong K sink and usually contains significant amounts of this element, with concentrations ranging from 0.55 to 0.80 mg/kg^{-1} fruit fresh weight (fw) (Zavalloni et al., 2001). The optimal K requirement ranges from 2 to 5% of plant dry matter, whether in vegetative parts, fruits, or tubers (Lopez, 2005). Aguilar et al., (2012), indicate that the application of higher doses of K avoids weight and water loss during postharvest storage. Foliar K is not always related to yield (Boonterm et al., 2010) because plants, in general, absorb more K than their metabolic needs and accumulate it in cellular organelles in luxury consumption. According to the review of Kant and Kafkafi (2018) it is observed that, if there is a high production of fruit or high grain yield, there is a higher consumption of K from leaves, stems and even roots, which is why when there is a demand for K from fruits, tubers or ears, the absorption of the nutrient is given by soil and other plant organs. High concentrations of K may be necessary during the early stages of growth to achieve maximum yields, with fruit of the appropriate size and optimum internal quality demanded by the markets.

Magnesium

In the case of magnesium it is considered indispensable in the formation processes of carbohydrates, oils and fats, it is essential for the function of various enzymes such as ATPases, RNA polymerases, protein kinases, phosphatases, glutathione synthetase and carboxylases, it helps to maintain homeostasis, since enzymes of the chloroplasts are affected by small variations in the levels of magnesium in the chloroplasts, and it is essential for the function of various enzymes such as ATPases, RNA polymerases, protein kinases, phosphatases, glutathione synthetase and carboxylases. element in the cytosol and chloroplast (Santos et al., 1999), its importance in plant growth and development, the molecular mechanism in plant cells that regulate homeostasis, and the relationship with transport within the plant, are poorly understood (Guo, 2016). In addition, Mg present in the vacuole is important for maintaining cation-anion balance and regulating cell turgor (Martinoia et al., 2000). In the leaf, 75% of the Mg content is involved in protein synthesis and 15-20% of the total is associated with chlorophyll pigments (White and Broadley, 2009). For this reason, this nutrient is required for the proper functioning of the numerous cellular activities related to chlorophyll synthesis, since it is the main component of this molecule, intervening as a cofactor of a series of enzymes involved in the metabolism and fixation of photosynthetic carbon, as well as participating in maintaining the stability of the membrane (Hermans et al, 2013). Regarding fruit quality, Lôpez et al. (2003) observed that the application of Mg to plums causes an increase in fruit firmness, but despite this, there is no clear evidence on the role of this element in the induction of increased firmness.

Calcium

Calcium favours cell division, stimulates hormonal and enzymatic processes, increases the speed of root development, can reduce germination, sporulation and the growth of pathogens, stabilises and ensures the permeability of the cell walls, protecting the fruit from degradation, which is why it determines its hardness and consistency (Saure, 1996; Garate, 2014), in addition, fruit resistance is related to the formation of calcium pectate in the cell membrane (Ramfrez et al., 2005). According to Barros (2016) proper Ca management delays ripening and improves storage qualities, as Ca treatments are also used to increase the postharvest life of a wide range of fruits and vegetables, These include: apple, strawberry, melon, pear, peach, tomato, mango and orange (Valdivezo, 2017). Galvis et al. (2003) found that direct post-harvest treatment of fruit with CaCl solutions$_2$ can be even more effective than pressure infiltration and can increase Ca concentrations of apples even more than vacuum infiltration, these methods of pre- and post-harvest treatment have inherent problems, as inadequate nutrient uptake is a problem in some cases because the fruit ingests an excess and injury occurs, or on the other hand not enough is applied to have a positive effect, according to Porro et al. (2006) based on a review of numerous studies of Ca deficiency factors, concluded that this phenomenon is the result of the failure of complex processes vital to the plant leading to physiological disorders in the fruit. Freitas and Mitcham (2012) found that the key role in the formation of the bitter pit physiological disorder could be played by Ca translocation from the peduncle to the calyx part of the fruit, a process that depends on the ability to bind Ca ions^{2+} in the cell wall terminal tissue of the fruit peduncle, the number of functional xylem tubes through which Ca^{2+} travels to the calyx, and the required hydrostatic concentration gradient. Based on the above, Joubert et al. (2008) cited that the incidence of bitter pit was halved when 'Cox's Orange Pippin' apples were immersed immediately after harvest in 0.25 mg $CaCl_2$, while Biskup et al. (2003) reported that in fruit sprayed with 3% CaO fertiliser, the lowest percentage of bitter pit was observed (12%), while the highest percentage was recorded in untreated fruit (39.64%). Another popular physiological disorder in apple, is the watery core, which is originated by the inadequate nutrition of the trees, especially regarding Ca and the presence of hot days alternated with cold nights, the intensity of this disorder is diminished in its severity if Ca is present in the fruit in sufficiently high quantities (Lopez, 2005), it is necessary to consider that the mineral status of Ca can involve other elements, especially K, P and B, (Mancera et al...), 2007), which is why it is important to properly manage the orchard to optimize the Ca uptake of the fruit in the pre-harvest environment to avoid losses due to lack of nutrients Ramfrez et al., (2005). To achieve an increase of Ca in fruits, foliar application of salts can be implemented during the growing season (Porro et al., 2006), for example, in 'Delicious' apples the fruit appearance, and to a lesser degree the red skin area, the control of internal browning, fruit firmness and to a lesser degree the acidity, were improved with foliar Ca sprays (Ramfrez et al., 2005), in this case the uptake of Ca and its distribution to the different organs increases as the rate of transpiration increases and, on the contrary, the inhibition of this transpiration decreases translocation, especially to the apical shoots and fruit, although it can also do so to the rest of the plant (Armstrong and Kirkby, 1979).

Sodium

Sodium is mobile within the plant and compared to other nutrients such as potassium and magnesium has a secondary significance within plant nutrition as it contributes to plant metabolism, photosynthesis and osmosis (movement of water in and out of plant cells) (Cortés, 2019).

MATERIALS AND METHODS

The research work was carried out in the 2020 cycle. It was carried out in the orchard and cold storage plant "La Campana", with the variety 'Golden Delicious' owned by Abram Olfert, located in the Mennonite field number 22 in the municipality of Cuauhtémoc, Chihuahua, Mexico, with an average altitude of 2048 meters above sea level, north latitude 28°26'17.5" and west longitude 106°53'40.3". Finishing and fruit quality analyses were carried out in the soil laboratory of the Facultad de Ciencias Agrotecnolôgicas of the Universidad Autônoma de Chihuahua.

Experimental Design

Completely randomised factorial experiment with four factors at four levels each. Bounded to 16 treatments generated using the Taguchi L16 experimental scheme. A linear and quadratic response surface was estimated by least squares regression using the SAS statistical package (SAS Institute Inc., SAS/STAT Software: Usage and Reference, Version 6, First Edition, Cary, NC: SAS Institute Inc., 1989). A factorial arrangement of 4 concentrations and 4 factors was used (table 2). The experiment was limited to 16 treatments in the Taguchi L16 structure (table 3) with four replications.

Table 2. Factors and levels of application of the Taguchi L16 structure.

Levels	Factors mM			
	K	Ca	Mg	Na
0.0	0.00	0.00	0.00	0.000
0.5	3.00	7.50	0.40	0.025
5.0	15.00	37.50	2.00	0.125
10.0	30.00	75.00	4.00	0.250
Simple average	15.00	37.50	2.00	0.125
Solution more than mM	100.00		5.00	0.25

Harvesting was carried out on 20 August, for the selection of 'GD' apple fruit at harvest, special care was taken in the field to obtain 10 commercial quality fruits per treatment, without physical damage or visible diseases. For each treatment, 5 L of water was added to a 20 L container, the amounts of stock solution indicated in table 3 were added and the solution was made up to 10 L by adding water and shaking. The fruits were immersed and shaken manually for 10 min. After this time, they were removed and left at room temperature for 10 min to drain off excess water. They were then placed in perforated plastic bags in groups for each treatment, to be stored in a controlled atmosphere for a period of 7 months. At the end of the process After storage, the fruits were taken to the laboratory and kept at room temperature to simulate shelf life. Five apples per treatment were evaluated for weight, colour, firmness, total soluble solids (TSS), titratable acidity (TA) and sugar-acidity ratio. The remaining fruits were used to determine the biological compounds: total phenols (TF) and antioxidant capacity (AC).

Table 3. Treatments formed in Taguchi L16 structure, application of mL of stock solution, for apple dips

Treaty	K	Ca	Mg	Na
1	0.00	0.00	0.00	0.000
2	0.00	7.50	0.40	0.025
3	0.00	37.50	2.00	0.125
4	0.00	75.00	4.00	0.250
5	3.00	0.00	0.40	0.125
6	3.00	7.50	0.00	0.250
7	3.00	37.50	4.00	0.000
8	3.00	75.00	2.00	0.025
9	15.00	0.00	2.00	0.250
10	15.00	7.50	4.00	0.125
11	15.00	37.50	0.00	0.025
12	15.00	75.00	0.40	0.000
13	30.00	0.00	4.00	0.025
14	30.00	7.50	2.00	0.000
15	30.00	37.50	0.40	0.250
16	30.00	75.00	0.00	0.125

Each treatment was diluted volumetrically in 10 L.

Fruit quality response variables

The following response variables were taken into account to assess the efficiency of the treatments:

Fruit finish

•Fruit weight. The weight was determined with an Ohaus ScoutTM Pro digital balance from 0 to 500.00 g.

•Colour. Two colour measurements were taken per fruit (intermediate sides in terms of colour) using the colour scale developed for 'Golden Delicious' by Soto et al, (2001) in six categories 1) green; 2) beginning of red striae formation; 3) uniform striae of opaque red colour; 4) striae of evident dark red colour; 5) less uniform striae, beginning of dark red colour; and 6) complete dark red, to make the scale more objective the colour was expressed as a percentage.

Fruit quality

•Firmness. The firmness of the fruit flesh was determined with a penetrometer (model Effe-Gi 327, 0-28 lb in2), two readings were taken for each fruit on the sides where the colour was measured and the individual average was obtained.

•Total soluble solids (TSS). The extract was obtained from two segments of each fruit, using a refractometer (Atago 0 - 32 °Brix) previously calibrated with distilled water.

•Titratable acidity. Ten mL of the juice extract was titrated, 6 drops of phenolphthalefin indicator (0.5 g phenolphthalefin plus 70 mL ethyl alcohol were added and made up to 100 mL with distilled water), titrated with a 0.1 N NaOH solution (2.15 g of NaOH 97 % purity, made up to 500 mL) until a brick pink-purple red colour was obtained; the volume used was converted as malic acid using the expression: % malic acid = ((((0,1* ml) / 10) * 67) / 10).

•Sugar to acidity ratio. It is expressed as parts of sugar to one part of malic acid (TSS/titratable acidity).

Bioactive Compounds

•Total phenols. They were determined according to the technique of Singleton and Rossi (1965), with slight modifications, using gallic acid as a standard. A quantity of 2 g of apple pulp was ground and extracted with 20 ml of 80% methanol. In a test tube 750 µl of 2% sodium, 250 µl Folin-Ciocalteau 50%, 1375 µl distilled water and 250 µl of the pulp extract. It was shaken and left to react for 60 min in the dark at room temperature.

The absorbance was measured at 725 nm in a DR 5000 Hach visible spectrophotometer. Results were expressed as g gallic acid per g fresh weight (g AG g^{-1}). A calibration curve was plotted. Linearity was determined between 0.5 and 2.0 mg ml^{-1} , using a high purity reagent grade gallic acid standard, calibration was measured in triplicate, the value of the equation was 6.2228x - 0.0107, with an r^2 of 0.9804.

• Antioxidant capacity. It was determined according to the methodology of Brand-Williams et al. (1995), with slight modifications. 2.8 ml of freshly prepared 0.1 mM DPPH solution (3.94 mg DPPH in 100 ml 80 % methanol) was added to a test tube, 0.2 ml of the supernatant of the homogenate used for the determination of total phenols was added, shaken in vortex and left to react for 60 min in the dark at room temperature. The absorbance was measured at 517 nm, using a DR 5000 Hach visible spectrophotometer. As a blank, the extract was replaced with 80% methanol and a capacity curve was plotted. Linearity was determined between 0 and 600 M using high purity reagent grade Trolox as standard, calibration was measured in triplicate. The equation had a value of 0.0008x + 0.6984 with an r^2 of 0.9855. The analyses were measured in triplicate. Results were expressed as mg Trolox g^{-1} fresh weight.

Statistical Analysis

The data obtained are subjected to a statistical analysis by linear and quadratic response surface to evaluate the significant variables. The analysis of each response variable comprised three stages: 1) analysis of the regression and the contribution of each factor to the regression fit; 2) canonical analysis of the response surface to determine the shape of the curve for those factors that had significant linear, quadratic and interaction responses; and 3) the values of each factor that had significant linear, quadratic and interaction responses. The predicted values depended on whether the minimum or maximum response was selected according to the original range of the data (Figueroa G, 2003). The behaviour of all response variables was summarised in a table where the factors and the simple average for each of them were specified. The resulting eigenvalues expressed as percentages of the mean are taken as positive or negative, as appropriate. The contribution of the eigenvectors was expressed with rounded signs so that 0,3750 :: ++ :: 0,6249, 0,6250 :: +++ :: 0,8749, ++++ >The same procedure was applied to the negative eigenvalues. In this way, the factors were weighted to determine which have the greatest influence on each variable.

RESULTS AND DISCUSSION

The data obtained from the physico-chemical evaluation of the fruits are presented in Table 4, where it is observed that the immersion of fruits in solutions of cations has a positive response with respect to the analysis of the response surface, since the factors A (K), B (Ca) and D (Mg) are above the value of the total frequency of eigenvectors which is 20, being A= 43, B= 41 and D=45, while the factor E (sodium)= 12, is below the selection. Therefore, the elements potassium, calcium and magnesium show influence in the experiment. For the canonical analysis of fruit immersion, those eigenvectors with a value equal to or greater than the variable (12.5% of the sum of the positive frequency), which was=14, were selected, all of them being selected: colour= 22, weight and firmness= 19, total soluble solids, titratable acidity and acidity sugar ratio= 17, phenol content= 16 and antioxidant capacity= 14, showing that all the variables are considered for a positive response of the experiment. For the weight variable, a mean of 135.4 g was obtained, while Mancera et al. (2007) reported a mean of 171.1 g for 'Golden Delicious' apples, which does not agree with our results. In the colour variable, Ornelas et al. (2018) compared 'Golden Delicious' apples with different harvests after flowering (DDF) and determined that at 122 DDF the colour is maintained with a value of 60.6 %, compared to this work there was a decrease of 5.3%, a very low value. considering that the fruit was subjected to a long storage period. In the case of firmness, Calvo (2002), carried out a postharvest experiment on Red Delicious apples with applications of 1-Methylcyclopropene (1-MCP), an ethylene action inhibitor, where he reports a mean of 7.2lb/in^2 for firmness, while in this work the mean was 9.8lb/in^2 , in addition, it also reported a mean of 10.8°B for TSS, the cation-based treatment reported a mean of 14.3 Brix, however, Flores (2018) reports 13.7 °Brix in apples treated with foliar calcium carbonate at harvest, which is very similar to the postharvest treated fruits of this work. In yellow and green apples, the TSS content must be higher than 12 % (NMX-FF-061-2003), which also agrees with our result. For the TA is known to be given by the organic acids present in apples, malic acid is the most important one in apples, although others such as succinic acid are also found. When the AT values are low during ripening, it is due to the degradation of the organic acids to sugars, decreasing the acidity content. In this work, 0.59% is reported, while Corona et al. (2020), reported 0.37 % malic acid in 'Golden Delicious' apples in postharvest, so this work had less degradation of organic acids.For the variable of sugar-acidity ratio, Oviedo et al, (2021), reported in their experiment of immersion with salicflico acid and nutrients a mean of 35.79, while in this work they reported a mean of 24.38, in this case, this parameter, provides the maturity index, in pomegranate is commonly used to define the "flavour" of the fruit during development, in addition, the sugar-acidity ratio increases as the fruit matures (Torres et al., 2022).

On the other hand, according to Asale et al, (2021), phenolic compounds are generally considered as a very important antioxidant source in fruits. In their work, the content of total phenols in the determination of 'Golden Delicious' apple fruit extracts using the Folin-Ciocalteu reagent were expressed in milligram gallic acid equivalents per gram (mg GGE g^{-1}), obtaining a value of 71.88 mg GGE g^{-1} p.f., while, Oviedo et al, (2021) report 480.14 g AG g^{-1} p.f., results that are more similar to ours, since

525.9 g AG g^{-1} p.f. were obtained, the variations may be due to the different methods of extraction and obtaining phenols. In the case of antioxidant capacity, the average we obtained was 3.5 mg Trolox g^{-1} p.f., Kupaeva and Kotenkova (2019), analysed Simirenko apples where they reported 5.41 µmol equiv. Trolox /gr, data very different from ours, both for the variety and the methodology used, on the other hand, Oviedo et al, (2021), report 3.05 mg Trolox g^{-1} of p.f., data that are similar to ours, data that are similar to ours, data that are similar to ours, but still below those reported in this work.Javaga (2019) conducted an experiment with different formulations of polyvinyl alcohol and potato starch to generate coatings, in order to demonstrate their effectiveness in the postharvest storage of apples of the 'Golden Delicious' variety, where it was determined that they had no notable effect on the metabolic physiology of the apple or on the mechanical properties of the fruit, probably due to its low viscosity, which resulted in a low surface density of the coatings used, discarding coatings of this type to treat postharvest fruit and implementing the use of immersions to extend shelf life, this work shows us the importance of cations, resulting with greater relevance of $MgSO_4$ [1.0] > K_2 SO_4 [1.0] > Ca [6.2] kg 1000 L water^{-1} . Observing that the exogenous application of $CaCl_2$ tends to decrease the permeability of cell membranes and consequently reduces water absorption, which contributes to the increase of fruit firmness and extends its shelf life, this once they are under refrigeration conditions (Raffo, 2000), in turn, Guerra et al. (2011) carried out sprays with calcium carbonate, to evaluate the effect on postharvest quality in Reinette de Canada' (RC) and 'Reinette Grise de Canada' (RG) apples, observing a higher incidence of bitter pit in 'RC' than in 'RG', due to the K/Ca ratio, therefore, calcium carbonate would be a useful product to decrease the incidence of bitter pit, but in this case the results may differ by apple variety as in our study there was no incidence of bitter pit, however, Joubert et al, (2008) used calcium nitrate ($Ca(NO_3)$), starting in three different varieties. different growth stages of 'Golden Delicious' fruit (early, medium and late). Late applications of $Ca(NO_3)$ (80 days after full bloom (ddfb)) increased the Ca content of the fruit at harvest more than early (six ddfb) and medium (40 ddfb) applications, generating a trend towards an increase in bitter pit from early to late applications of $Ca(NO_3)$, demonstrating that calcium applications are necessary in the 'Golden Delicious' variety due to its susceptibility to bitter pit and if it is in the late ripening stage of the fruit, it has a high influence on the quality of the fruit. Also the interaction between cations and their application at harvest can be an important factor to avoid physiological disorders and to improve shelf life. Accordingly, Krishkov (2007) cited that the degree of bitter pit appearance is associated with K and Mg concentration, independently of Ca content, and Almeselmani et al. (2010) suggest that an addition of K in the fertilisation of tomato plants can help to preserve fruit during postharvest storage, in this sense it is shown that immersion of apple fruit in K solutions$^+$ improves fruit shelf life, This is in agreement with the work of Shen, et al (2019), where foliar applications of K and its antagonistic effect with Mg and Ca in pear trees, showed that the interaction that was affected was that of Mg, as the concentrations were low in the leaves, However, the concentrations of K, Ca and Mg in the fruit increased, from this we realise that the interaction of these elements in the fruit does not generate antagonism as long as it is applied in a balanced way, comparing with our experiment it can be observed that the use of these elements in immersion had a positive influence, perhaps because of the synergy that is generated in the fruit and helped the preservation in controlled atmosphere. In the case of Mg, it has been described that foliar application of this nutrient to peaches and nectarines produces an increase in firmness at harvest, but this effect does not persist during postharvest (Serrano et al., 2004), so its efficiency in this experiment can be attributed to the different combinations with the other cations and to the crop. For example, a postharvest study of Anna apples dipped in $MgCl_2$ and $CaCl_2$ alone or in combinations provided evidence that $CaCl_2$ combined with $MgCl_2$ have the highest firmness of apple fruits, furthermore, the highest TSS values were found in fruits treated with $CaSO_4$ and $MgSO_4$ compared to untreated fruits (Kiram, 2012), according to this work, the combination of Mg, K and Ca cations help to maintain the firmness and TSS content of the fruit.

According to Monge et al, (1994) immersion of apple fruit for 10 minutes in a solution of 0.125 mg/l of $CaCl_2$ seems to be more advisable than $Ca(NO\ )_{32}$, due to the risk of nitrate ion residues remaining in the apples, and improves fruit quality. This supports our work, since applications of $CaSO_4$ (6 kg/1000 L of water) notably improved the appearance of fruit treated in post-harvest in the same way. On the other hand, Conway and Sams (1995) tested the technique of infiltration by vacuum or under pressure with different CaCl $solutions_2$ in 'Golden Delicious' apples, finding that the ideal treatment was infiltration under pressure with a 4% solution, but this method can cause problems in apples with lesions as the solution can penetrate into the core of the fruit and cause rotting, these techniques are more costly and risky compared to the application of cation solutions directly to the fruit.

Table 4. Canonical correlation for 'Golden Delicious' apples post-harvest 2020. harvest time

Factores / media simple (L 1000 L^{-1} agua)						
	K_2SO_4	Ca	$MgSO_4$	NaCl		
	(A)	(B)	(D)	(E)		
Eigenvalores[U]	**2.8**[T]	**7.5**	**0.8**	**9.3**	Eigenvectores	
	Peso de fruto (μ 135.4; 95.0 – 140.0 g)[W] C[Y]				Subtotal	+ / -
168.1(143.0)	-3(+2)[V]	(-3)	+3(+2)		13	7/6
-62.1	+2	+2	+2		6	6/0
|Freo.|/ resp.	7 C	5 C, A	7 C, B		19[X]	13/6
Dosis	4.0	4.0	0.2	9.3	Selección= 3	
Color (μ 55.3; 46.0 – 60.0%) C, P						
44.3(32.9)	+3(-3)	+3(+2)	-3(+2)		16	10/6
-16.8	+2	+2	+2		6	6/0
	8	7 **	7 **		22	16/6
	2.0	1.0	1.0	8.0	Selección= 3	
Firmeza (μ 9.8; 9.0 – 12.5 lb in^2) C, P						
32.1		+3	+3	+3	9	9/0
-65.2(-39.2)	+4	(-2)	(-2)	(+2)	10	6/4
	4 L, C	5 L	5 L, C	5 L, C, A, B, D	19	15/4
	1.0**	6.2**	0.99**	9.71**	Selección= 2	
Sólidos solubles totales SST (μ 14.3; 15.0 – 18.0 °Brix) L, C, Pr						
128.0(99.5)	(-3)	+3	(-2)+2		10	5/5
-80.4	+3	+2	+2		7	7/0
	6 L, C	5 L, C, A	6 L, C, A, B		17	12/5
	2.0**	9.3**	0.16**	9.0	Selección= 3	
Acidez titulable (μ 0.59; 0.4 – 0.7 % ácido málico) C, P						
193.2(172.9)	-3	(-3)	+3(+2)		11	5/6
-79.66	+2	+2	+2		6	6/0
	5 C	5 C, A	7 C, A, B		17	11/6
	4.0**	3.0**	0.3**	9.4.	Selección= 2	
Relación azúcar acidez (μ 24.38; 23.0 – 32.0 SST / % ácido málico)						
27.1	+3	+3		+3	9	9/0
-65.6(-44.5)		(-2)	-2(+2)	(+2)	8	4/4
	3	5	4	5	17	13/4
	4.0	4.0	0.5	9.4	Selección= 3	
Contenido de fenoles (μ 525.9; 468.8 – 619.2 μg ácido gálico g^{-1} p.f.) L, C, Pr						
29.0	+2	+2	+2		6	6/0
-55.3(-40.8)	-2(+2)	-3	(-3)		10	2/8
	6 B, E	5 C,D	5	L	16	8/8
	2.9**	4.2**	0.5**	9.5**	Selección= 3	

[T]Simple mean of each factor;[U] Eigenvalues expressed as a percentage of the mean, values in brackets correspond to the second eigenvalue and their respective eigenvectors.[V] each sign counts from the second quartile onwards (+2 from 0.375 - 0.624, +3 from 0.625 -0.874, +4 > 0.875, same for negative signs); μ overall mean, XInterval of the response variable estimated by the regression,[X] Observed frequency for that variable, multiplied by 15% (-1) of the total sum of eigenvalues to be considered; Significant factors* (0.05 :: Pr :: 0.01), highly significant** (Prob < 0.01);[Y] significant regression linear (L), quadratic (C), and interactions between factors (A, B, D, E). [Z]Total frequency of eigenvalues for the set of variables, selecting those factors whose sum is equal to or greater than 20% of the total positive eigenvectors.

The figures below show the variables that were most significant with the contribution of cations.

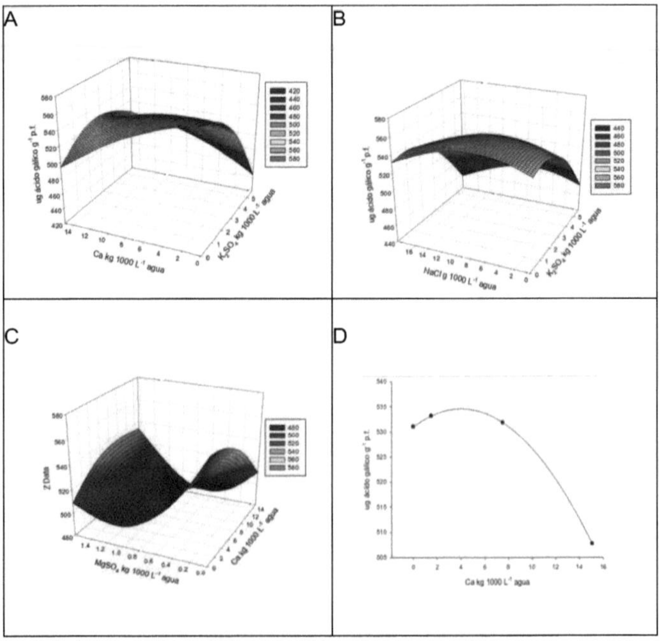

Figure 1. Phenolic compounds with respect to cation balance.

Figure 1 shows the behaviour of the phenolic compounds in relation to the cations. In section A, the interaction between Ca and K is observed, where individually both elements in high concentrations have a negative effect on the concentration of TF, in this case K produces the greatest effect due to the antagonistic actions between these two elements, however, joint applications of 2-6 kg of Ca and 1-2 of K can maintain the phenol content up to 580 g of AG g^{-1} of p.f. In section B is the interaction between Na and K, where the joint applications of 1-2 kg of K and 7-12 g Na can be maintained up to 580 g AG g^{-1} p.f., above these amounts the total phenol content can drop to 440 g AG g^{-1} p.f. The behaviour of Ca individually (C) shows that as the Ca content increases in the immersion, the TF falls below 510 g AG g^{-1} p.f., while inputs of 4 kg of Ca give a maximum of 534 g AG g^{-1} p.f. Finally, the interaction between Mg and Ca shows antagonism between the elements. In fact, individual applications of Mg tend to have negative effects on phenols, but if they are combined with high amounts of Ca they tend to conserve the amounts of total phenols as long as they do not exceed 10 kg of Ca inputs, even so, Oviedo et al., (2021), report an average of 434.09 g AG g^{-1} p.f., in 'Golden Delicious' apples 13 days after harvest, so our work shows good results. In this case, Ca and K are important in the water balance, Ca forms part of the cell membrane and is stored between the cell wall and the intermediate lamina, where it interacts with peptidic acid to form calcium pectate, providing stability for its integrity. Equally important, it is involved in the regulation of enzyme systems and phytohormone activity, increasing tissue resistance to pathogens, as well as postharvest shelf life and nutritional quality (Yfran et al., 2017). Similar work to ours using different combinations of treatments on apple found a significant increase in firmness when combining $CaCl_2$ (1% and 2%) and $MgCl_2$ (1%

and 2%) salts to maintain higher tissue firmness (Farag and Nagy, 2012).

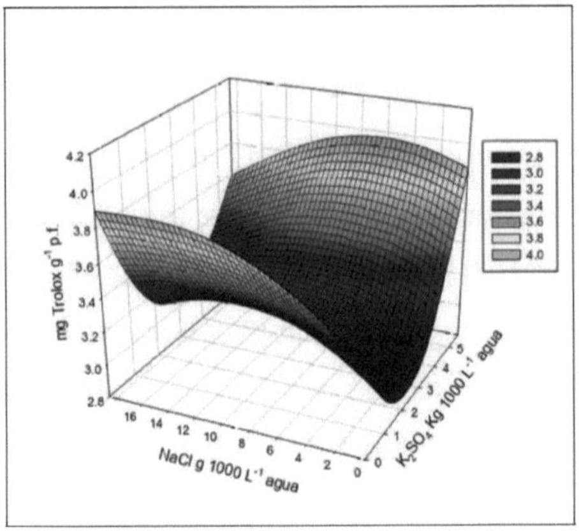

Figure 2. Antioxidant capacity with respect to Na and K

In Figure 2 we can observe the interaction between sodium and potassium, where both elements applied in their maximum concentrations tend to increase the antioxidant capacity (AC), in turn, with a contribution of 6-8 g of Na and 5 kg of K the AC can be increased to 4.0 g of Trolox g of p.f., In this case, K, individually, has a negative tendency at low concentrations, unlike Na, which tends to increase the AC. Oviedo et al. 2021, obtained 4.59 mg trolox g-1 p.f. with concentrations of 1,320 mM of K, which is very similar to the data obtained in this work.

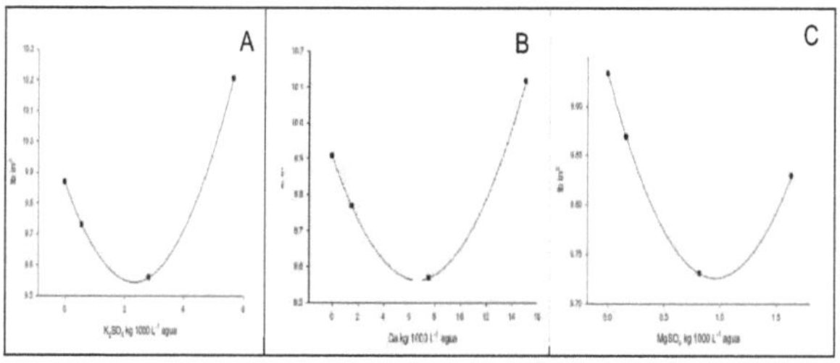

Figure 3. Firmness with respect to cation balance

For firmness (Figure 3), K and Ca showed similar behaviours, where maximum inputs tend to increase apple firmness, applications of 5-6 kg of K and 12-15 kg of Ca maintain firmness up to 10.2 lb in^2 . In the case of Mg, 0-0.2 kg of this nutrient preserved firmness above 9.85 lb in^2 . In a study conducted on peach, it was reported that immersions in Ca salts at 62.5 mM maintained fruit firmness during storage (Manganaris et al., 2007), on the other hand, a study conducted on 'Golden Delicious' apples reported an average of 10.05 lb in^2 at 13 days after harvest (Oviedo et al., 2021), our work managed to maintain firmness in 10.2 lb in^2 after refrigeration for 7 months.

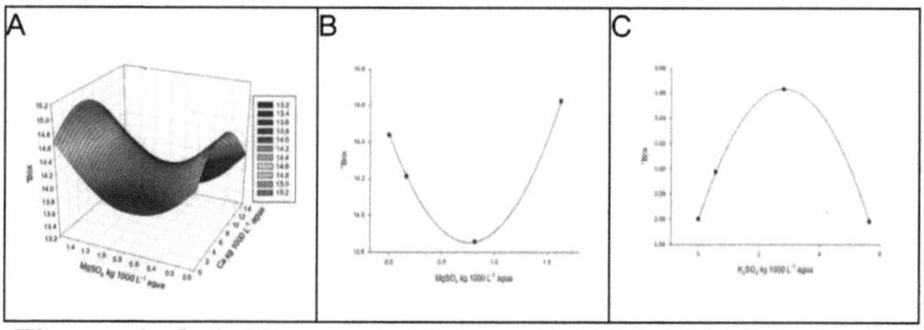

Figure 4. Soluble solids with respect to cation balance

In the case of TSS (Figure 4) inputs of 1.5 kg of Mg and 6-8 kg of Ca can maintain the concentration up to 15.2 °Brix, in this case we can also observe (A) that average inputs of both elements will result in the lowest TSS content, as we also observe in the section B. For K (C), an input of 3 kg can contribute to increase 0.66 °Brix, Oviedo et al. 2021, mention in their experiment a mean of 13.8 °Brix, while in this work the mean was 14.3 °Brix, and it may be due to the different interactions in the treatments.In the case of acidity the interactions between the cations are shown in Figure 5, the synergistic interaction between Ca and K (A) to maintain the malic acid content, at concentrations of 4-8 kg Ca and 2-4 K up to 0.68 % acidity can be maintained. Individually, neither of the two elements promotes the malic acid content. In section B, a maximum percentage of 0.61% is observed with a maximum of 4 kg Ca, and above this amount, the acidity drops. In section C, the interaction between Mg and K is observed again with a maximum of 0.61%. synergism, where contributions of 2-4 kg of k and 0.2-1.4 kg of mg can obtain an acidity of up to 0.66%, a behaviour that cannot be observed individually. For the interaction between Mg and Ca (C) a linear behaviour is shown, where the joint application only goes down to 0.58% and the highest acidity is obtained with the lowest calcium and the highest Mg inputs, obtaining 0.62%. Finally, the individual behaviour of Mg (E) is observed; the incorporation of 1.5 kg of Mg in the immersion maintains the acidity up to 0.59%. Guerra et al (2011), in their postharvest experiment with apples treated in immersion with $CaCl_2$ at a concentration of 2%, for 30 s, after 60 days of cold storage, served to retain the acidity of the fruit until 120 days of storage, which agrees with our experiment.

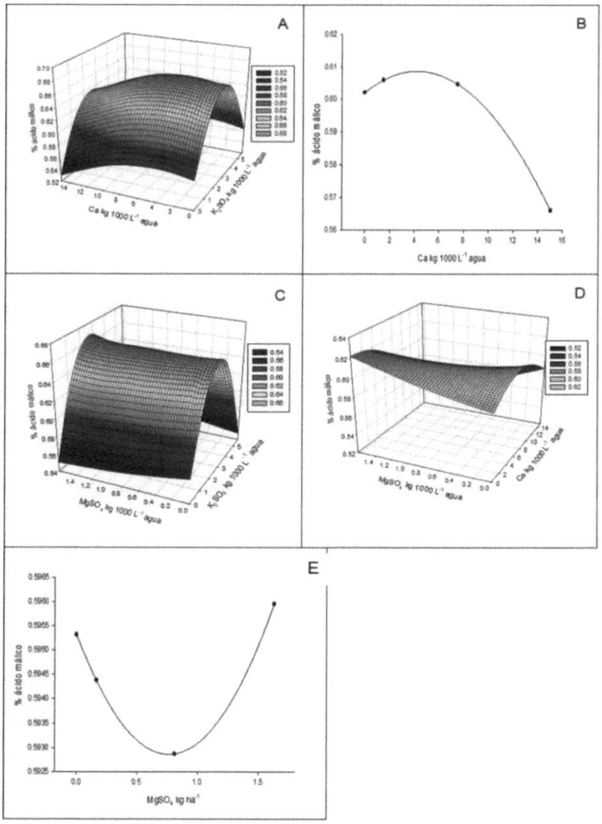

Figure 5. Percentage of acidity with respect to cation balance

CONCLUSIONS

The positive effect of immersion in cation solutions on 'Golden Delicious' apples had a positive postharvest effect with a dose of $MgSO_4$ [1.0] > K_2SO_4 [1.0] > $CaSO_4$ [6.2] kg 1000 L water^{-1} , however, $NaCl_2$ showed no relevant results in the experiment, as it had no influence on the parameters evaluated. In the work, Mg was the cation that showed the most influence, followed by K and finally Ca. Contrary to what is believed that only Ca is indispensable to maintain fruit quality, there are more elements that participate in the process. It was observed that all the variables showed a response to the treatments, first the colour, followed by the firmness, which shows that in spite of the time the fruit is subjected to refrigeration, there is no great loss of weight, while the sugar-acidity ratio (TSS / % malic acid) is maintained in optimal ranges, a quality that makes the fruit acceptable to the consumer. Finally, the phenol content and antioxidant capacity reflect how the balance of cations has an influence on them, since depending on the dosage, increases or decreases of both are presented, improving the contribution of bioactive compounds of the fruit. The need to develop commercially acceptable methods to sustain the postharvest life of fruits has led to the development of different technologies, so that the immersion of fruits in cation solutions can be a functional alternative for the preservation of fruit in cold storage for long periods of time, in order to increase the shelf life of apples.

LITERATURE CITED

Aguilar C., Barrameda Y., Montesinos B., Romero T., Soriano J. M. (2012). Effect of potassium biofortification on cherry tomato postharvest: implication of some phenols. nutriplanta.

Almeselmani M., Pant R. C., Shing B. (2010). Potassium level and physiological response and fruit quality in hydroponically grown tomato. January, International Journal of Vegetable Science.

Arellano J. (2021). Characterisation of native apple varieties to determine their added value for fresh consumption. academica- e.unavarra.es.

Armstrong M. J. and Kirkby E. A. (1979). The influence of humidity on the mineral composition of tomato plants with special reference to Ca distribution. Plant and Soil. 52: 427-435.

Asale V. Dessalegn E. Assefa D. Abdisa M. (2021). Phytochemicals and antioxidant activity of different apple cultivars grown in south ethiopia: case of the wolayta zone. Pages 354-363.

Bai J., Hagenmaier R. D., Baldwin E. A. (2002). Volatile response of four apple varieties with different coatings during marketing at room temperature. Journal of Agricultural and Food Chemistry, 50, 7660-7668.

Barraquilla, A. (2005) The apple tree. Biologia: http://encina.pntic.mec.es/~esarment/biologia/imagenes/Manzanweb.pdf

Barrôs M. (2016). Technical guidance for apple tree cultivation. University of Santiago de Compostela. Santiago: LEADER.

Ben J. (1998). Estimation of apple storage quality based on fruit analysis. Int Sem: Ecological Aspects of Nutrition and Alternatives for Herbicides in Horticulture. Warszawa, Poland, pp: 7-8.

Biskup S., éosié T., Pecina M., Miljkovié M. (2003). Utjecaj folijarne gnojidbe kalcijem na njegov sadrzaj u plodu jabuke (Effect of calcium spray on calium conet in apple fruit). Pomologia Croatica: Glasilo.

Blackhall V. Curetti M, Colavita G. (2018). Magnesium deficiency in apple trees. F&D Magazine N° 81 - 1°. INTA. https://inta.gob.ar/documentos/deficiencia-de- magnesium-in-apple tree.

Boonterm C. W., Silapapun A., Boonkerd N. (2010). Effects of nitrogen, potassium fertilizer, and clusters per vine on yield and anthocyanin content in Cabernet Sauvignon grape, Suranaree. Journal of Science and Technology, Peshawar, v.17, p.155-163, 2010.

Calvo G. (2002). Effect of 1-Methylcyclopropene (1-MCP) on CV. Red delicious apples harvested at three stages of maturity and preserved in conventional cold and controlled atmosphere RIA. Research Journal Agropecuarias, vol. 31, nüm. 3, December, pp. 9-24. Instituto Nacional de Tecnologfa Agropecuaria Buenos Aires, Argentina

Candan A. P. and Calvo G. (2017). Dynamic controlled atmospheres: an alternative for the control of surface scald in pears. RIA, 42, 291-299.

Castillo, J. (2018). Chihuahua smells like apples. El Souvenir, pp. https://elsouvenir.com/chihuahua-huele-a-manzana/.

Cepeda I. Saucedo C. Colinas M. T. and Rodrfguez J. (2014). Evaluation of pre- and postharvest treatments with cacl2 on the cold storage and quality of golden Delicious apple. Colegio de Postgraduados Pregep-fruticultura.

Conway W. and Sams E. (1995). Relationship between total and cell wall bound calcium in apples following postharvest pressure infiltration of calcium chloride. Acta Horticulturae 398, pp 31-39.

Corona L. S., Hernandez D., Meza O. (2020). Analisis de parametros fisicoqufmicos, compuestos fenôlicos y capacidad antioxidante en piel, pulpa y fruto entero de cinco cultivares de manzana (malus domestica) cosechadas en México. Instituto Politécnico Nacional. Escuela Nacional de Ciencias Biolôgicas- Zacatenco. Instituto Politécnico Nacional, Escuela Nacional de Ciencias Biolôgicas-Santo Tomas.

Cortés, E. C. (2019). Manual manejo de manzano. Facultad de Agronomfa de la Universidad Nacional (sede Medellfn), pp2.

Falcon J (2016). New trends for the treatment of apple bitter pit: Reineta, a case study. Departamento de Nutriciôn Vegetal Estaciôn Experimental de Aula Dei Consejo Superior de Investigaciones Cientfficas (CSIC).

FAO. (2010). Fertilizers and their use. Parfs: IFA. Retrieved from http://www.fao.org/3/a-x4781s.pdf

Ferguson B. and Watkins B. C. (2003). Cation distribution and balance in apple fruit in relation to calcium treatments for bitter pit. Scientia Horticulturae. https://www.sciencedirect.com/science/article/abs/pii/030442388390078X

Fernandez D. Bautista S. Ocampo A. Garcfa A. Falcôn A. Rodrfguez V. (2015) Edible pellicles and coatings: a favourable alternative in the postharvest preservation of fruits and vegetables. Revista Ciencias Técnicas Agropecuarias. Universidad Agraria de La Habana, Facultad de Ciencias Técnicas, San José de las Lajas, Mayabeque, Cuba. Instituto Politécnico Nacional-Centro de Desarrollo de Productos Biôticos, Yautepec, Morelos, México. Instituto Politécnico Nacional,

Escuela Superior de Ingenierfa Mecanica y Eléctrica, Unidad Zacatenco Secciôn de Estudios de Posgrado e Investigaciôn, Mexico City, Mexico. Instituto Nacional de Ciencias Agrfcolas (INCA), Departamento de Fisiologfa y Bioqufmica Vegetal, San José de las Lajas, Mayabeque, Cuba.

Figueroa G. (2003). Optimization Of A Response Surface Using Jmp In. Mosaicos Matematicos No. 11. Departamento de Matematicas Universidad de Sonora.

Flores M. A., Soto J. M., Salas A., Sanchez E., Pifia F.J. (2018). Effect of $CaCO_3$ industrial byproduct on quality atributes, phenolic content and antioxidant capacity of apple cvs Golden Delicious and Top Red. Nova scientia Departamento de Nutriciôn Vegetal, Universidad Autonôma de Chihuahua. Tecnologfa de Alimentos y Productos Lacteos, Centro de Investigaciôn en Alimentaciôn y Desarrollo A.C., Unidad Delicias. https://doi.org/10.21640/ns.v10i20.1190

Freitas S and Mitcham E (2012). Factors involved in fruit calcium deficiency disorders.
Hortic Rev 40: 107-146. http://dx.doi.org/10.1002/9781118351871.ch3

Galvis J. A., Arjona H., Fischer G. (2003). Effects of the application of calcium chloride (CaCl2) solutions on storage life and fruit quality of mango (Mangifera indica L.) Van Dyke variety. Agronomfa Colombiana, vol. 21, nüm. 3, pp. 190-197.

Garate M . (2014). "Apple production. Pp 1-82. http://saludpublica.bvsp.org.bo/cc/bo40.1/documentos/704.pdf.

Gomila, T. (2016). New tools for post-harvest optimisation in the region. Revista Fruticultura & Diversificaciôn.

Guerra M., Marcelo V., Valenciano J. B., Casquero PA. (2011). Effect of organic treatments with calcium carbonate and bio-activator on quality of 'Reinette' apple cultivars. Sci Hortic129 (2): 171-175. http://dx.doi.org/10.1016/j.scienta.2011.03.013

Guo W., Nazimc H., Liang Z., Yangab D. (2016). Magnesium deficiency in plants: An urgent problem. The Crop Journal. Volume 4, Issue 2, pp. 83-9. https://www.sciencedirect.com/science/article/pii/S221451411500121X

Gutierrez A. C., Soto J.M., Sanchez E., Yanez R. M., Flores B. (2014). Macronutrient fertilization in apple trees: yield and foliar micronutrient content. Revista biolôgico agropecuaria Tuxpan.

Hermans C., Conn S. J., Chen J., Xiao Q., Verbruggen N. (2013). An update on magnesium homeostasis mechanisms in plants Metallomics, pp. 1170-1183.

Hernandez E. (2010). Pepper (capsicum annuum l. cv. darsena) nutrition as a function of the ratio of no_3 / nh_4 and K concentration in the nutrient solution under greenhouse conditions. Centro De Investigacion En Qufmica Aplicada.

Jackson D. and J. Palmer (1999). Temperate and subtropical fruit production. Pome fruits. pp. 189-202. In: 2nd ed. CABI Publishing, Wallingford, UK.

Javaga A., Sapper M., Martfn M., Gonzalez C. (2019). Effect of the application of coatings based on biopolymers and carvacrol on postharvest quality of apples

https://riunet.upv.es/bitstream/handle/10251/124751/J%C3%A1vaga%20-%20Efecto%20de%20la%20aplicaci%C3%B3n%20de%20recubrimientos

%20a%20base%20de%20biopol%C3%ADmeros%20y%20carvacrol%20sobre%20la%20ca pdf?sequence=1&isAllowed=y

Joubert J., Lotze E., Theron K. I. (2008). Evaluating pre-harvest foliar calcium applications to increase fruit calcium and reduce bitter pit in 'Golden Delicious' apples. Sci Hortic 116 (3): 299-304. http://dx.doi.org/10.1016/j.scienta.2008.01.006

Kant S., and Kafkafi U. (2018). Potassium uptake by crops at different physiological stages. University of Jerusalem, V, 38. Retrieved from https://www.ipipotash.org/uploads/udocs/Sesion V.pdf

Karim M. (2012). Effect of Pre- and Post-Harvest Calcium and Magnesium Compounds and Their Combination Treatments on "Anna" Apple Fruit Quality and Shelf Life. Journal of Horticultural Science & Ornamental Plants 4 (2): 155- 168.

Koch M., Busse M., Naumann M., Jakli B., Smit I., Cakmak I., Hermans C., Pawelzik E. (2019). Differential effects of varied potassium and magnesium nutrition on production and partitioning of photoassimilates in potato plants Physiol. Plant.,166, pp. 921-935.

Krishkov E. (2007). Factors influencing the incidence of bitter pit on apple fruits and control measures. Proc Agric Sci XL (2): 22-26.

Kumar R., Khurana A., Sharma A. K. (2014). Role of plant hormones and their interplay in development and ripening of fleshy fruits. Journal of Experimental Botany, 65, 4561-4575.

Kupaeva N. and Kotenkova E. (2016). Search for alternative sources of natural plant antioxidants for food industry. Gorbatov Federal Research Center for Food Systems of Russian Academy of Sciences, Moscow,

Russia. Dmitry Mendeleev University of Chemical Technology of Russia, Moscow, Russia.

Kyanko M. V., Russo M., Fernandez M., Pose G. (2010) Effectiveness of peracetic acid on the reduction of the spore load of postharvest rotting moulds of fruits and vegetables. National University of Quilmes, Argentina.

Leôn G. (2016). Systems of production vegetal https://www.uaeh.edu.mx/investigacion/productos/4781/sistemas_de_pro ducci on_vegetal_2.pdf

Lester G., John L., Donald J., Makus D. (2010). Impact of potassium nutrition on postharvest fruit quality: Melon (Cucumis melo L) case study. Plant Soil. DOI 10.1007/s11104-009-0227-3

Liu H. and Han Z. (1997). Apple fruit mineral nutrition. J Fruit Sci14 (z1): 73-78.

Lopez C., Botia M., Alcaraz C. F., Riquelme F. (2003). Effects of foliar sprays containing calcium, magnesium and titanium on plum (Prunus domestica L) fruit quality. Journal of Plant Physiology 160, 1441-1446.

Lopez, M. (2005). Nutritional characterisation and quality of 'Red Delicious' and 'Golden Delicious' apples from two producing countries. Chihuahua: Autonomous University of Chihuahua.

Mancera M. M., Soto J. M., Sanchez E., Yafiez R. M., Montes F., Balandran R. (2007). Mineral characterisation of 'Red Delicious' and 'Golden Delicious' apples from two producing countries. TECNOCIENCIA Chihuahua, 1(2), 6-17. https://doi.org/10.54167/tecnociencia.v1i2.42

Martinoia E., Massonneau A., Frangne N. (2000). Transport processes of solutes across the vacuolar membrane of higher plants. Plant and Cell

Physiology 41, 1175- 1186.

Monge E., Val J., Saenz M., Blanco A., Montafiez L. (1994). Calcium as a plant nutrient. Bitter pit in apple trees. Department of Plant Nutrition (E.M., J.V., M.S., L.M.) and Department of Pomology (A.B.), Estaciôn Experimental de Aula Dei (C.S.I.C.).

Morales, M. (1999). In Variedades y calidad de las manzanas de Aragon (p. p.). 70). Zaragoza: Apeph.

Musacchi S. and Serra S. (2018). Apple fruit quality: overview on pre-harvest factors. Department of Horticulture, Tree Fruit and Research Extension Center (TFREC), Washington State University, 1100 N. Western Avenue, Wenatchee, WA 98801, USA.

Ornelas J. J., Quintana B., Escalante P., Hernandez J., Pérez J., Rios C., Ruiz S. (2018). Relationship between the firmness between Golden Delicious apples and the physicochemical characteristics of the fruits and their pectin during development and ripening. J Food Sci Technol. Jan; 55(1): 33-41.

Oviedo J. C., Soto J. M., Sanchez E, Yafiez R. M., Pérez R., Noperi L. C. (2021). Salicylic acid and nutrient immersion to maintain apple quality and bioactive compounds in postharvest 49(3). https://doi.org/10.15835/nbha49312409

Palomo I., Yuri J. A., Carrasco R., Quilodran A., Neira A., (2010) El Consumo De Manzanas Contribuye A Prevenir El Desarrollo De Enfermedades Cardiovasculares Y Cancer: Antecedentes Epidemiolôgicos Y Mecanismos De Acciôn. Revista Chilena de Nutriciôn, vol. 37, nüm. 3. pp. 377-385 Sociedad Chilena de Nutriciôn, Bromatologfa y Toxicologfa Santiago, Chile.

Pavicié N. (1993). Predicting the occurrence of physiological disorder bitter pit in fruits of Golden Delicious and Idared apples. Agronomski Glasnik 52: 419-425.

Ponce O., Soto J. M., Noperi L., Alvarez A. Ochoa J., Holgufn F. (2020). Production and quality of Golden Delicious apple by effect of macronutrient fertilization. Instituto del Nacional de Investigaciones Forestales, Agrfcolas y Pecuarias (INIFAP). La Campana Experimental Field. Aldama, Chihuahua, Mexico. Autonomous University of Chihuahua. Faculty of Agrotechnological Sciences.

Porro D., Ceschini A., Pantezzi T. (2006). The importance of advisory service in predicting bitter pit using early-season fruit analysis. Acta Hort 721: 273-277. http://dx.doi.org/10.17660/ActaHortic.2006.721.37.

Posadas M., Lôpez P. A., Gutiérrez N., Dfaz R. and Ibafiez A. (2018). Phenotypic Diversity Of Apple In Zacatlan, Puebla, Mexico Is Broad And Is Given Mainly By Fruit Traits Breni. Scielo 2018).

Raffo D. (2000). Calcium applications and fruit quality. Revista Espafiola INTA 1: 10-15.

Ramfrez J. M., Galvis J. A., Fischer G. (2005). Postharvest ripening of feijoa (Acca sellowiana Berg) treated with $CaCl_2$ at three storage temperatures. Agron. colomb. vol.23 no.1 Bogota Jan.

Ruales P. (2019). Determination of soil cation exchange capacity and its correlation with the exchangeable cation content of plants of the genus Siparuna. http://dspace.utpl.edu.ec/handle/20.500.11962/24597.

Santos T., Aguilar A., Manjarrez D. (1999). Foliar fertilization, an important support for crop yields. Mexican Society of Soil Science, A.C. Chapingo, Mexico, vol. 17 nüm. 3 pp. 247- 255.

https://www.redalyc.org/pdf/573/57317309.pdf

Saure M. (1996). Reassessment of the role of calcium in development of bitter pit in apple. Funct Plant Biol, 23(3),237-243. http://dx.doi.org/10.1071/pp9960237.

Sanchez E. and Curetti M. (2021). Fruticultura Nutriciôn De Las Plantas Zona Templada, Suelo y Aplicaciôn Abonos. Ediciones INTA, Estaciôn Experimental Agropecuaria Alto Valle. Mineral nutrition of temperate fruit trees.

Serrano M., Martfnez D., Castillo S., Guillén F., Valero D. (2004). Effect of preharvest sprays containing calcium, magnesium and titanium on the quality of peaches and nectarines at harvest and during postharvest storage. Journal of the Science of Food and Agriculture 84, 1270-1276.

Shen Ch., Shia X., Xiea Ch., LiaHan Y., Xinlan Y., Yangchun M., Caixia D. (2019). The change in microstructure of petioles and peduncles and transporter gene expression by potassium influences the distribution of nutrients and sugars in pear leaves and fruit. Journal of Plant Physiology Volume 232, January 2019, Pages 320-333

Soto J. M, Pifia M, Pifia F. J, Sanchez E, Pérez R. and Basurto M. (2007). Fertigation with macronutrients in 'Golden Delicious' apple trees: impact on yield and fruit quality. redalyc.org/pdf/2033/203345704010.pdf

Tagliavini M., Brunetto G., Wellington J., Bastos De Melo M., Maurizio T., Quartieri (2015). The role of mineral nutrition on yields and fruit quality in grapevine, pear and apple Bras. Frutic. 37. https://doi.org/10.1590/0100-2945-103/15

Tarrillo L. (2017). "Disefio de un prototipo de un secador homogéneo de frutas utilizando flujo controlado de aire caliente". Undergraduate thesis,

Universidad Catôlica Santo Toribio de Mogrovejo.

Torres N. G., Soto J. M., Sanchez E., Noperi L. C., Yafiez R. M., Pérez R. (2022). Comparison between quality and bioactive compounds of pomegranate from two producing areas. Notulae Scientia Biologicae, 14(2), 11267-11267.

Tovar B., Mata M., Garcfa H. S., Montalvo E. (2011). Effect Of Wax Emulsions And 1-Methylcyclopropene On The Postharvest Preservation Of Guanabana. Revista Chapingo Serie Horticultura.

Trainotti L., Tadiello A., Casadoro G. (2007). The involvement of auxin in the ripening of climacteric fruits comes of age: the hormone plays a role of its own and has an intense interplay with ethylene in ripening peaches. Journal of Experimental Botany, 58, 3299-3308.

Valdivezo I. (2018). "Postharvest application of calcium chloride on apple fruit (Malus x domestica Borkh) cv. ANNA" Aula Dei Experimental Station. Dei EEAD CSIC.http://repositorio.lamolina.edu.pe/bitstream/handle/20.500.12996/3477/ valdiviezo-aliaga-ivan-abel.pdf?sequence=3

Valdivezo J. (2017). Calcium, determinant in the postharvest performance of stone and pome fruit. Estaciôn Experimental de Aula Dei EEAD - CSIC. https://digital.csic.es/bitstream/10261/165239/1/ValJ_%20I-EncuentrPoscosech-1_2017.pdf

White M. and Broadley R. (2009). Biofortification of crops with seven mineral elements often lacking in human diets: iron, zinc, copper, calcium, magnesium, selenium and iodine. New Phytol., 182, pp. 49-84.

Xie K., Cakmakc I., Wanga S., Zhang F., Shiwei G. (2021). Synergistic and antagonistic interactions between potassium and magnesium in higher plants The Crop Journal Volume 9, Pages 249-256. https://doi.org/10.1016/j.cj.2020.10.005

Zaera, P. (2008). Fruit quality evaluation in apple trees. Retrieved from http://digital.csic.es/bitstream/10261/18601/1/Proyecto%20Pilar%20Dolz.pdf

Zavalloni C., Marangoni M., Scudellari D., Tagliavini M. (2001). Dynamics of uptake of calcium, potassium and magnesium into apple fruit in a high density planting orchard. Proc. 5th Intl. Symp. Mineral Nutrition of Deciduous Fruit Crops. Penticton Canada.

MIX
Papier aus verantwortungsvollen Quellen
Paper from responsible sources
FSC® C105338

Printed by Books on Demand GmbH, Norderstedt / Germany